# SEWERAGE

AND

# SEWAGE UTILIZATION,

BY

PROF. W. H. CORFIELD, M. A.

OF THE UNIVERSITY OF LONDON.
AUTHOR OF "WATER AND WATER SUPPLY."

NEW YORK:
D. VAN NOSTRAND, PUBLISHER,
23 MURRAY AND 27 WARREN STREET.
1875.

# PREFACE.

This essay is slightly condensed from the Lectures of Prof. Corfield, delivered before the School of Military Engineering, Chatham, England.

The rapidly increasing importance of the subject demanded the republication in this country of whatever was original in England, by so eminent an authority. The Lectures first appeared here in the pages of Van Nostrand's Magazine, from which this little book is carefully reprinted.

# SEWERAGE

AND

# SEWAGE UTILIZATION.

---

The water is brought into the town to be soiled, and it must be removed ; and besides this dirty water which has to be removed, there is the surface water, and the subsoil water that have to be removed also ; together with a quantity of refuse of all sorts, with various impurities from manufactories, from slaughter-houses, from animal sheds, together with slops from private houses, and so on. This impure water is carried away from towns by means of pipes, known as sewers, and I want at once to explain to you in a few words, the difference that is to be kept in sight between a sewer and a drain. A sewer is a pipe for removing impure water, water that has been fouled ; a drain, as Mr. Bailey Denton said in a letter to the *Times*, is meant to take the wetness out of soil ; it is meant to dry the soil—it is not meant to carry away impure water.

As these sewers are to carry away impure water, it is perfectly plain they must be impervious to water, or they may, on certain occasions, let it leak out into the subsoil of the town underneath the houses, and also into the wells, if there are any. If they are impervious, the water of the soil, the subsoil water at any rate, won't get into them and so they will not act as drains. Now you will see directly why it is necessary to drain the subsoil underneath the streets and houses. That it is necessary I can show you in a half a minute.

It has been perfectly clearly shown by Dr. Buchanan, from statistics of the death-rate of certain towns that have been sewered, that in those towns which have had sewers so constructed that the subsoil water of the town has been lowered, the death-rate from consumption has increased in a most extraordinary manner. In the case of certain towns the death-rate from consumption has been reduced by half the total number of deaths, by 50 per cent, by the lowering of the subsoil water consequent upon sewering the town as it is called. But these sewers were so constructed that they acted as drains as well. Towns which have been sewered with impervious pipes throughout, so that no reduction of the subsoil water had been effected have shown no decrease in the death rate from consumption, and some have shown an increase. So that shows you that it is necessary to drain the subsoil.

Then from the incompatibility of having pipes which both drain the subsoil and are impervious, so as not to allow of the sewage to escape from them, it has been suggested to have two systems—to have drains and sewers. Mr. Menzies has been the great advocate of having what is called the separate system. His plan was to have deep sewers, pipe sewers, which are impervious, and then rather superficial drains to carry off the flood waters. This plan would not provide for actually draining the subsoil unless some special provision were made for it. The usual plan that is practised is to build sewers large enough to contain all the drainage water and any reasonable amount of storm water that may fall upon the land which is sewered, but it is perfectly ridiculous to use them for intercepting natural watercourses, as is done in so many cases.

Some sewers in the South of London actually collect water from natural watercourses which ought to be allowed to run straight into the Thames. The argument for admitting this extra amount of water into sewers is that they will be kept cleaner, and that they will flush themselves naturally, as it were. But against this is to be placed the difficulty of dealing with the increased amount of water at the outfall. I may speak to you of that, however, bye and bye.

Now for laying main sewers you must

have accurate plans of the places, with the levels of the surface along the roads and the streets, and the levels of the deepest cellars, so that the sewer of the street may always be below the level of the lowest cellers. You must also know the levels of high and low tides, if near to the sea. The general plan, according to Mr. Rawlinson, ought to be made on a scale of two feet to a mile, and the detailed plan on one of ten feet to a mile.

I am now going to refer you to an important discussion that took place before the Institution of Civil Engineers in 1862 and 1863. You will find it in Vol. 22 of the *Proceedings of the Institution of Civil Engineers.* I shall have to refer to this discussion several times. The first point to be attended to in laying out main sewers is that they shall be straight from point to point. There is no reason that they should follow exactly the middle of the streets where they are not straight, but they should be made straight from one point to the next. The curves should be gentle, not greater than $22\frac{1}{2}$ degrees, for instance. The junctions should be, as in the case of the main water pipes, curved. Rankine tells us that main sewers should not be less than two feet broad, and that the velocity in them should not be less than one foot in a second, for fear of choking up, nor greater than four feet and a half in a second, because with a greater velocity than this you have too much scouring. The usual plan, then, is to make these drain sewers, as I call them, sewers which are capable to a certain extent of acting as drains also. That end is often realized by setting the bricks of the invert, as it is termed, in cement, and setting the others with mortar. The bricks of sewers ought always to be set in hydraulic mortar or in cement. These drain sewers, I should tell you, are on the plan of the oldest sewer we know of, namely the Cloaca Maxima in Rome. That Cloaca Maxima was not constructed as a sewer : it was originally a drain. A great deal of blame has been thrown upon the Romans because the Cloaca Maxima was not made impervious ; but we must remember it was originally constructed as a drain. It was formed to drain off the water about the Forum, and it did so, and does so to this day. It only came afterwards to be used as a sewer, that is to say, to have refuse matter thrown into it, and that is no doubt how we have got our system of drain sewers, and there is no doubt that the sewers in many towns in England were originally built as drains.

The first thing to mention is the trench. Mr. Rawlinson tells us, in a paper that I have already quoted to you, and which is entitled "Suggestions as to plans for main sewerage and drainage," that the most difficult earth to deal with is quicksand, and as a rule it should only be opened in short lengths. The trench may require to be close timbered ; and in all cases the greatest care should be exercised in taking the timbers from the sides of the trenches so that none of the side earth may fall down upon the sewers.

With regard to the depth of the trenches, of course this must vary very much in different places. The only condition is that they require to be placed deep enough to drain all the cellars. I may mention, as an example, that at Stratford-on-Avon the sewers are constructed from 16 feet deep down to 4 or 5 feet in many other parts. At Rugby they average 11 feet in depth, but they vary from 7 to 25 feet. One of those papers that I quoted to you from the *Proceedings of the Institution of Civil Engineers* (Vol. xxii., p. 265), says, that the average depth is 12 feet, but that the depths vary much. Tunneling may be required, as practised now in the large outfall sewer being constructed at Brighton. Tunneling, however, should never be resorted to when it can be helped, because much better supervision can be exercised over the construction of a sewer when you have a trench than if you have a tunnel. This is perfectly clear, and also for the same reason, night work should not be encouraged: the men should work in the day time.

Now as to the incline. The incline, we are generally told, should not be less than 1 in 600. Sometimes, of course, that cannot be got. The incline must vary very much with the natural incline of the soil. If possible, you should have it about 1 in 600 in mains, and a greater incline in the smaller sewers, and the greatest incline in the house sewers. The incline of the pipe-sewers that come from the houses should not be less than 1 in 60. Where sewers are joined the incline should be greater. Where a small sewer enters into a large one there should be a quicker incline for some little distance.

Another point is this—that a larger sewer should never open into a smaller one; neither should a sewer open into one of the same size, but always a smaller one into a larger one. The inverts should not be level. The invert of a smaller one should be higher up than that of the larger one, so that there may be a fall. "Main sewers and drains should be adapted," as Mr. Rawlinson says, "to the town area, length of streets, number of houses, surface area of house yards and roofs, number of street gullies, and volume of water supply."

With regard to the size and shape of the main sewers. The size is, of course, very variable indeed. I told you that Professor Rankine said they should not be less than 2 feet broad. They are often made less than two feet broad. Perhaps the best thing is to give you an example. It is taken from a discussion in the volume of the *Proceedings of the Institution of Civil Engineers*, which I referred to a few moments back. Mr. Newton said that "In purely urban districts a rainfall of one inch in half an hour ought to be provided for ; thus, on the 29th July, 1857, he registered at Preston three quarters of an inch of rain in 35 minutes, and on the 8th October, 1861, nearly the same depth in 30 minutes. On the latter occasion an egg-shaped brick sewer, 4 feet 9 inches high by 3 feet 2 inches wide, and 300 yards in length, with a fall of 1 in 156, carried away this water from a closely built and densely populated district containing 117 acres. In another part of the town, which was also built upon, and which contained 85 acres, a sewer 3 feet 6 inches high by 2 feet 4 inches wide, with a fall of 1 in 75, carried off these storms without causing any damage, and without the water rising in the cellars, which were generally from 1 foot to 2 feet below the soffit of the arches. In both cases, however, the sewers were under pressure, and on the first occasion the water rose 18 inches, and in the other case 1 foot, in the man-hole shafts." (Vol. xxii., p. 295.)

"The London main sewers vary from 4 feet in diameter, to 9 feet 6 inches by 12 feet in some cases. The three northern outfall sewers are each 9 feet by 9 feet with vertical sides, the southern outfall sewer 11 feet 6 inches in diameter." It is a good plan to make what are called intercepting sewers if there are considerably different levels in the town, or if the sewage has to be pumped at the outfall. This you know is done with the sewage of London on both sides of the river. There are two intercepting sewers in the south of London, and the sewage runs by gravitation, in the high level sewer, right away to the outfall at Crossness, and by the southern sewer, it runs also by gravitation, as far as Greenwich, where it is all pumped up into the outfall sewer, and then runs away to Crossness, where it is all pumped up into the Thames.

On the north side of London, there are three, and the sewage of the two lower ones is pumped up into the highest at Abbey Mills, and thence flows on to the outfall at Barking Creek.

Now for the shape. The best shape has been decided to be the egg shaped section. There are plenty of shapes in use. The rectangular section is evidently bad. The amount of friction is very great and likewise such sewers become choked up with deposit. A flat top has been used, but it is obviously bad. It is not so strong ; and even the Romans, as in the Cloaca Maxima, used an arch. The best shape is an oval section with the smaller end downwards. Another advantage of this is that there is a saving of a material. Sewers less than 2 feet in diameter are better made circular. There was, for a long time, a dispute as to the different advantages of brick and pipe sewers. You will find in the 12th Vol. of the *Proceedings of the Institution of Civil Engineers* a paper, a very important paper, by Mr. Rawlinson, in which he supported very strongly the use of pipes. You will find that there was a great deal of dispute as to their efficacy. Mr. Rawlinson laid down three propositions that he thought should be borne in mind in laying out the sewerage of a town. In the first place, the sewers cannot receive the excessive flood water even of the urban portion of the site. That is perfectly true; they have in certain places been made large enough to do that. In the second place, according to Mr. Rawlinson, they ought not to be combined with the natural watercourses which drain large areas of the suburban land previous to entering the urban portion. No doubt sewers are frequently combined with watercourses which ought to go directly into the rivers. In the third place, they should be adapted exclusively to carry the liquid and solid refuse from the houses in such a manner as to cause the least possible nuisance to the inhabitants. These conclusions were then very much disputed, as also was the conclusion that sewers should be as small as possible and impervious.—The opponents, no doubt, who disputed these statements, did so from the fact that they did not sufficiently appreciate the antagonism that exists between sewers and drains. Mr. Robert Stephenson, on that occasion, expressed his "conviction that for certain localities. if pipe-drains were sufficiently strong to resist fracture, and sufficiently large to avoid being choked up, they might be advantageously employed to form the connections of houses, courts, and other small localities, with the main sewers, which should be constructed of brick, of such dimensions as to admit of easy internal inspection and repair, and be of form (except where the flow of water was at all times considerable) that the radius of the curved bottom should be able to gather a small supply of water into a sectional area affording the same hydraulic mean depth as in a pipe-drain of a diameter merely adapted to discharge the minimum flow." So that after all this discussion the result which was come to was this: that impervious pipes—glazed earthenware pipes—were, on the whole, the best for house drains, small streets, courts and places of that sort, but that they were not advantageously to be used over 12, 15, or at the most 18 inches in diameter. Certainly, when above 18 inches in diameter, it is cheaper to make a brick sewer of oval section than to lay pipes.

In very wet soil, Mr. Rawlinson has used iron inverts to prevent the subsoil water coming into the sewer and keeping it continually full up to a certain height. Mr. Simpson has described iron pipes to be used for sewers where there are bad foundations, as in running sand; there is a plan for preventing subsoil water from getting into sewers without using cast-iron pipes, which is described by Messrs, Reid and Goddison, of Liverpool, in the British Association Report for 1870. They have introduced a subsoil drain and pipe rest to be placed beneath the pipes. It has got a section like the letter D. The pipe sewer is laid upon it, and it acts as a drain to keep the subsoil water below the sewer.

With regard to the outfalls—the outfalls ought, if possible, to be quite free. The first thing that you have to do is to choose the best place for the outfall. For that there are no general rules whatever, and you must be guided entirely by the nature of the locality. Most sewers having been originally constructed as drains, it is perfectly plain that their only outfall is into the sea or into a river, and so most of the outfalls are built into the sea or into rivers, and the sewage is thrown away. We shall in the next lectures consider some other methods of dealing with sewage.

If possible the outlet should be free. If it cannot be free there must be some means adopted for preventing the sewage getting backed up in the sewers, especially when the rivers are high, or at high tide in the sea. If the sewage is allowed to get back in the sewers you may get the cellars flooded, and you will certainly get sewer air forced up into the town. One way of preventing this is by causing the outfall to open into a large tank out of which the sewage is continually pumped. Another plan is simply to have a flap to the mouth of the sewer—a flap which shuts and keeps it full of sewage. In that case the outfall has to be made large enough to contain an enormous quantity of sewage. Then it should certainly not be taken—that is if you drain into a river—into a river near the town, and certainly not above one. The better method is perhaps to have a large tank, if the outfall sewer must be below the surface of the water. Where you have rivers with considerable difference in the level, a plan has been adopted for discharging the sewage in summer when the river is very low by means of a subsidiary pipe. This cast-iron pipe is taken at a lower level into the river. There is a valve capable of being raised by a windlass, which valve prevents the sewage coming out by the main outfall. The ordinary sewage of the town can then run away by this cast-iron pipe, and get off into the river at a lower level. This plan has been put into operation at Windsor by Mr. Rawlinson; so that the ordinary amount of sewage need not run out by the main outfall high up when the river is low, in which case it would run down the banks causing a nuisance. When there is an enormous amount of sewage and a flood in the river, and of course the river is high, then it is allowed to come out by the main outfall. When there are steep gradients in sewers there ought to be steps made, and flaps placed at the upper parts, and at such places also there ought to be ventilators. Ventilators are best constructed to open at the level of the street, and are best made in connection with man-holes. In the first place I ought to tell you that it is absolutely necessary to ventilate sewers. It is perfectly certain that a certain amount of sewer air, as it is called, is contained in all sewers, and is given out from sewers, and is given out from sewage whether there is much stagnation or not. Where there is great stagnation the more is evolved. The strongest argument for ventilating sewers is the argument used by those who say they should not be ventilated. They say they should not be ventilated because they can be securely trapped, and the small amount of gas that does accumulate in them can be prevented from coming into the houses. Now this fact that sewers need to be trapped is the best argument to show that it is neccessary to ventilate them. I should tell you that all water traps are, essentially, bends in pipes which will hold water. Water traps are of very little use against sewer air. I do not mean to say that the air will often actually force them, though it will do that sometimes ; but what I mean is that most of the dangerous elements which are the constituents of sewer air are soluble in water and are evaporated and given out, so that it is very little use to rely on water traps, especially if they are placed under pressure. Therefore ventilation must be provided.

Now this ventilation has been carried out in very various ways. The simplest way, of course, is to allow a certain number of the openings into the sewers in the streets to be untrapped, and then the sewer air escapes into the streets.

That was the plan condemned some years ago, because it allowed the sewer air to come out straight into the streets, and to become disagreeable to persons walking. That plan, however, is certainly very much better than letting it remain in the sewers, from which it will get into the houses, as it is certain to do, through weak points.

Another plan was to have special ventilating pipes carried up to the top of high buildings. These are very well in their way, but they are certainly not sufficient. Sometimes at the top of these pipes, Archimedean screws have been placed. At Liverpool an enormous quantity of these Archimedean screws have been placed at the top of such pipes, and more are now being placed. But I must tell you with regard to these that Drs. Parkes and Burdon Sanderson, in their Report lately on the Sanitary Condition of Liverpool, made experiments, and found that these Archimedean screws altered the pressure of the air in the sewers to a very trifling extent, so that they did not seem to be of any great value.

Another plan is to connect the rain-water pipes with the sewers directly, or at least some of them, and to leave them untrapped. If you do this, it is necessary that rain water pipes should be thoroughly well constructed and well jointed, or else air will escape into the neighborhood of the houses. But the best plan of all is to have plenty of openings into the main sewer directly over it, and to have special ventilating openings connected with the man holes along each sewer at certain intervals, in the middle of the streets. You must have man holes, and where a sewer makes a bend there ought to be a man hole; and there likewise there ought to be a ventilating shaft, and also at every one of the steps which I formerly mentioned to you. Where you have a steep incline you should have a step and a fall, and there, there ought to be a flap and a ventilation shaft provided with charcoal trays.

There are two or three ways of doing this, one is to have a ventilating shaft at the side of the man hole. The air that comes up the man hole passes into the ventilating shaft and through the charcoal and out into the street, and the air is deodorized by passing through the charcoal. That is one way. The dust and dirt which will collect at the bottom of the shaft can be easily removed through the man hole. Another plan is to suspend charcoal trays at intervals in the man hole itself, and then to have an opening through which the deodorized gas goes. If you have plenty of these openings along the sewers into the streets there will not be much nuisance. Then besides ventilation, sewers generally require flushing, or at any rate cleaning out. A deposit occurs in certain parts. Now the old plan used to be to make all main sewers so that a man could go through them. That plan is not now employed, because flushing has been adopted instead of cleansing out by hand labor, that is to say to a very considerable extent. Flushing is performed either by stopping the sewage at certain places and so giving it a higher head, which is the plan often adopted, or by having some special reservoirs of water (collected for the purpose) at the higher parts of the sewers which can be allowed to rush down them; or again, by making arrangements with the water companies for the supply of a sufficient amount of water to flush them continually. And they should be flushed regularly, or deposit is sure to occur. The Paris plan of flushing the sewers is interesting. You know they have in Paris enormous subways under the streets, and the sewer runs along at the bottom of the subway. This subway has a rail on each side of it, and they flush the sewer in this way: a wagon is run along these rails, and there is a flap which descends from the wagon into the sewage below. The force of the sewage pushes this flap on, and carries the wagon on too, and the flap of course displaces everything before it. A certain amount of space is left beside the flap, so that the sewage rushes past this, and it in fact chases everything before it that stands in its way, so far as deposit is concerned. Of course the expense of flushing sewers with water is very much less than that of cleansing them by hand labor. I could give you some instances of the amounts of the cost in each instance, but I do not know that it is necessary.

We have now followed the course of water from the place where it is collected into the town, and we have also described sewers. I began by describing to you the outfalls, because that is the natural way of proceeding, not because the water followed that course, but because, before you have small drains and sewers in a town, you want the outfall and the main sewers.

I have now before I go any further a few more points to tell you with regard to the house sewers, or house drains, as they are generally called. In the first place, I have already told you that the fall of the house drains should not be less than 1 in 60. Then, the next point is that house drains ought not to run underneath the basements of houses. They generally do so, as you know perfectly well. If they do they ought to be made of impervious pipes laid in concrete. The next point is that they ought invariably to be ventilated. If the water closet system is used perhaps the best way of ventilating them is to allow the pipe which comes from the closets—the soil pipe—provided it descends outside the house, as it always should, to be untrapped at the bottom, and to be open at the top, so that the air from the house sewer finds exit into the open air continually. If the water closet plan is not adopted there ought to be one or more special pipes for ventilating the drain carried up to the highest point. Or, again, some of the rain water pipes can be left untrapped; but this is not so good a plan. If the soil pipe be inside the house it should be trapped at the bottom and ventilated at the top, and then a special ventilating pipe must be provided for the sewer. Another plan, an excellent one, in addition to this, if the house drain be long enough, is to cut it off—to make a break in it, as it were—before entering the main sewer, and that is done by making it discharge into a ventilating shaft. There is a swing flap on the end of the house drain in the shaft, which is shut except when the water is running. The air which comes up from the street sewer into the shaft cannot pass up the house drain, but ascends through trays of charcoal, and finds its way out into the open air through openings which are left between the bricks at the top of the shaft. The whole thing is covered with a stone slab, just above the level of the ground. If you want additional security you can place a syphon between the ventilating shaft and the street sewer.

Now if a trap is placed in the cellars, or basement of a house communicating with the drain, a precaution has to be taken, if there is any chance of the sewage backing up. In that case a trap has to be placed which will prevent sewage from flooding the basement. This is done by means of a heavy flap trap.

The common traps used for yards, and even for back kitchens, and so on, are what are called bell traps. They are about the worst kind of things that could be devised, and that is why I mention them. The bell trap merely consists of a sort of inverted tumbler placed over the head of the pipe that leads into the drain. The rim of this tumbler dips into a groove, which is supposed to be filled with water. The water that passes through the perforated top which is fixed on to the tumbler can find its way round the edges of this bell, as it is called, and so into the drain. The danger is this, that the instant this top is taken off it takes the bell off as well, and then sewer air can get up into the house or yard. Now these things are continually being taken off, or left off, and therefore that kind of trap should never be used. The best to put instead of it is an earthenware syphon trap. The advantage of this is, that if the top is taken off, as it continually is, to sweep the yard or basement, it does not matter at all, bacause the top has nothing to do with the trap itself. With this syphon, if you want to ventilate the drain at that particular point, you can have a hole made at the top of the bend (some are made with a hole), and then you can carry up a ventilating pipe from it. If you have two ventilating pipes you must carry them to different heights, one not very high, and the other to a considerable height; but practically one is sufficient. Another thing that you can do—especially if the trap is in the basement of the house—you can make any waste pipe end in the side of it, through a hole in the side above the water, and yet below the cover, so that you do not get the place flooded if the holes in the cover are stopped up, as they are apt sometimes to be. This is a very convenient plan.

Sinks ought always to be against external walls. They almost always used to be built (and now often are) against internal walls. Their pipes have no more business to go straight into drains than the waste pipes of cisterns; they should always be carried out into the yard, and made to end over one of these traps, or else they should be carried into the side of it. The same thing is true of rain water pipes. Unless rain water pipes are constructed with the view of ventilating the sewer, and are made with proper joints, they ought to end *above* the traps.

We now come to the consideration of the disposal of a particular kind of refuse matter, namely, excretal refuse matter. I want first to prove to you that it is necessary to get rid of refuse matter generally, and especially so of this particular kind of refuse matter from the neighborhood of habitations. I could quote to you from any number of reports showing that the general death rate, and also the death rate from certain specific diseases, especially typhoid fever and cholera, depends to a very great extent upon the amount of filth, and especially of excretal filth, that is in and about the habitations of people.

Take the following opinion from the evidence given by Mr. Kelsey before the Health of Towns Commission (1844). When asked, "Does the state of filth and the effluvia caused by defective sewerage, by cesspools or privies, and decomposing refuse kept in dust bins, powerfully affect the health of the population?" he says, "Yes, it does; it always occasions a state of depression that renders persons more liable to be acted upon by other poisons, even if it be not the actual cause of it. The line of habitations badly cleansed, and in this condition, *almost formed the line of cholera cases.*"

Then, after a description of cellar dwellings, which are even now prevalent in some of our large towns—in this case referring to Liverpool—Dr. Duncan pointed out that the ward "where the largest proportion (more than one half) of the population resides in courts or cellars, is also the ward in which fever is most prevalent, 1 in 27 of the inhabitants having been annually attended by dispensaries alone;" and he remarks that "people do not die simply because they inhabit places called courts or cellars, but because their dwellings are so constructed as to prevent proper ventilation, and because *they are surrounded with filth*, and because they are crowded together in such numbers as to poison the air which they breathe."

Well, then, illness is caused if these refuse matters are not removed from the neighborhood of habitations, and illness with all its attendant misfortunes and difficulties.

Now what plans have been adopted for removing these matters from habitations? That is one thing to be considered; and another thing to be considered is, are these excretal matters of any value; can anything be done with them; and if so, how can the most be got out of them?

Now, if I tell you what their composition is, you will see at once that they must be of considerable value; and when you reflect that these refuse matters constitute a great proportion of the refuse matters of our bodies, you will see at once that they must contain the same elements as our food, and that therefore there is at any rate a possibility of their being used for the reproduction of food. Now, what is their composition? The results of a great number of analyses, which are, however, only sufficiently complete in the case of males of from 15 to 50 years of age, show that the mean amounts in ounces of the various constituents during 24 hours are as follows:

| | Fresh Excrements. | Dry Substance | Mineral Matter | Carbon. | Nitrogen. | Phosphates |
|---|---|---|---|---|---|---|
| Fæces....... | 4.17 | 1.041 | 0.116 | 0.443 | 0.053 | 0.068 |
| Urine....... | 46.01 | 1.735 | 0.527 | 0.539 | 0.478 | 0.189 |
| Total..... | 50.18 | 2.776 | 0.643 | 0.982 | 0.531 | 0.257 |

[Paper read by Mr. Lawes, F. R. S., before the Society of Arts, March 7th, 1855.]

Now, what does that mean? Nitrogen and phosphates are the very things we get to use as manures, and we can from this chemical composition of the excreta calculate their relative and absolute value.

In the first place I want to point out to you that the amount of valuable matter contained in the urine in the 24 hours is considerably greater than that contained in the fæces in 24 hours. Now, that I tell you at once, in order that you may not run away with the fallacy that is sometimes indulged in that you may throw away the urine of a population so long as you retain the fæces, and that you will get the greatest amount of manure from the latter. On all heads the matters contained in the urine are in larger proportion than in the fæces, and especially as regards the important matters, *e. g.*, the nitrogen is about nine times as much.

To estimate the value, it is convenient to take amounts that are passed in a year, and it has been calculated that the average amount of ammonia—representing the nitrogen in the form of ammonia —discharged annually by one individual, taking the average of both sexes and of all ages, is about 13 lbs., or nearly that; and it has been estimated that the money value of the total constituents of the excreta is, in urine, 7s. 3d., and in fæces about 1s. 3d., giving a total of about 8s. 6d. a year, so that you see at once that the value of the urine is about six times as much as the value of the fæces. When you consider that about ten times more urine is passed (by weight) than fæces, you see that fæces are more valuable than urine, weight for weight, although the total fæces are much less valuable than the total urine. There is, then, no doubt about the value.

The next thing is, what are the plans that have been attempted for utilizing it? The earliest plan, and one that is defended by many up to the present day—and by many, I was going to say, who ought to know better—the earliest plan consists in keeping the fæces, and a certain small amount of urine for a longer or shorter period in or about the premises in some form or another. And there are two ways in which this can be done. It can be done, as it is in many towns even now, as it is notably in many continental towns, as Paris and Berlin and Vienna. It can be done either by keeping these matters in a semi-liquid state, in tanks or vessels prepared to receive them, and emptying these at certain times and taking their contents away to be used as manure, or it can be done by mixing these matters with certain refuse which will to some extent dry them ; and some such refuse is found in all houses, and is, to wit, ashes. Now, those are the two plans that have been adopted—I may almost say from time immemorial, at all events for a great many years—in order to collect this valuable manure; that is to say, by those who have made any attempt to collect it at all.

Let us take the first plan and consider it for a few minutes—the plan of digging a hole in the ground and throwing all this refuse matter into it. When this is done, unless the hole in the ground is impervious, a great amount of this refuse matter will percolate into the soil around, and get into wells. In certain towns this has been actually encouraged. There are certain towns where holes have been made to receive the refuse matters of the population in pervious sandstone strata, with the express, distinct, and avowed object of letting the liquid matters, and as much as possible of the solid matters, precolate the soil and get away as best they could. These dumb wells, as they are called, have been made and shut up with the deliberate intention of not being opened for many years, and in certain places the soil has been so absorbent that when opened the wells have been almost invariably found empty. Now that plan need only be stated to be condemned. In all these towns the well water, which is often the only supply for the people, is largely polluted, and is in fact to a great extent supplied by these very dumb wells, which are often close by; and in almost every town where there is an epidemic of typhoid fever you find an inspector going down from the Local Government Board, and reporting that this is the case.

Now, the improvement on this bad plan, or want of plan, in places where it is not done away with, is to line the pits with cement, and to provide a drain from them into the nearest sewer. Thus the cistern becomes merely a pit in which to collect the solid matters, while you allow the liquid matters, which are the most valuable, and which are just as likely to become offensive, to run into the sewer. You collect the solid matter which is less valuable, and which is rendered still less valuable by having much of its valuable material dissolved out, by the liquid which is allowed to run away.

The other plan is to do as is done in Paris, to make these large cesspools (so large that they take six months, or even a year, to fill), under the houses or under the courts, to make them impervious, and not drain them at all. Of course the pits are only theoretically impervious, but practically very many of them certainly are not so. But, however, supposing that they are, they in any case requiring a ventilating shaft, or the foul air which collects in them will find its way through, somehow or other, and will poison the air of the house. Another danger is, that if they are not ventilated, and even sometimes if they are ventilated, the men who go into them may be suffocated by the poisonous gases accumulated in them.

The first of these plans, in which the liquid matters all run away, is confessedly a failure. You deliberately take and throw away all the most valuable part, and the part which remains is not only of no value, but is a distinct expense, because no one will take it away unless he is well paid for it, so that there is a very considerable loss on that system. And so the system cannot be called one of utilization. Then, the disadvantage of the Paris plan, apart from the general disadvantage of having such a thing as an immense cesspool underneath each house, is found in the emptying of them. This operation causes a fearful nuisance, even although they are now emptied by means of carts in which a partial vacuum is first created, so that when the hose is attached to the cart and placed into the pit the semi-liquid stuff rises up and fills the tonneau, as they call it. And then I may tell you, as a matter of fact—I could give you the figures—that the system does not pay; that the collection costs so much that the manure made from the stuff does not pay the cost of collecting.

Let us consider, now, some improved systems in which this manure is collected, mixed with ashes and household refuse, and sold in a semi-dry state, because this is a plan which has been very much defended of late. These plans are developments of the old midden—a heap in the yard at the back of the house, into which all kinds of refuse were thrown.

The first improvement, as in the case of the cesspool, was to make a kind of pit lined with cement. There are different contrivances employed. There is one which is known as the Manchester plan, and another known as the Hull plan, and so on; but in all the ashes are thrown into the pit, so as to make a semi-solid mass.

The conditions necessary for them are these:—In the first place they must receive no moisture from the soil around. In the second place, they ought to allow no liquid to escape from them, because, if so, they are confessedly failures. It is plain that the object of all these systems where the excretal matters are to be kept out of the sewers must be to separate them *entirely* from the sewage, because, if you do not, you have still sewage to treat. We have already seen that the water supply of the town goes into it to be soiled, and you require sewers to take it away. You have got, therefore, water which is, to a certain extent, dirty. The theory of persons who support those systems which I am now describing is that, if you prevent the most foul part of the refuse matters from getting to the sewers, you will not then require sewers so large to begin with; and, also, that you will not require to treat the sewage afterwards, but will be able to turn it into a river without any disadvantage. Now, if these cesspools and these midden closets require to drain their liquid contents into the sewers, the sewage will certainly have to be treated just as much as if you were to allow the whole of the refuse matter of the town to get into the sewers; and that is proved by the fact that the sewage of towns where you have cesspools and midden pits drained into the sewers is considerably more foul, and is within a very little of being as strong, as the sewage of water closeted towns, so that it requires treatment at least as much as the latter does.

The next condition with these midden heaps is that they must secure, practically, as much dryness of the contents as possible; and then they require an efficient covering up of the refuse matters by the ashes that are thrown down, and that is done in various ways. Lastly, they require ventilating.

Now, with either of these systems, it is desirable to have the receptacle as small as possible ; it is desirable for sanitary reasons, though not for economical ones, to have the receptacle as small as possible, so that as little of these matters shall be retained about the premises as possible. The midden closet used at Hull consists of an impervious receptacle, which is not sunk into the ground at all, but is, in fact, merely the space directly under the seat of the closet, the front board being movable, so that the scavengers can get the stuff when full. That, no doubt, is by far the best of these simple ash closets.

After this we pass on to a still greater improvement, to what you might call a temporary cesspool—that is to say, a simple tub or box placed underneath the seat to collect the excretal matters, these tubs or boxes being collected every day by contractors and their contents used for manure.

I may tell you at once that this is really the system out of which most is got in the way of profit. There is no doubt of it whatever. This is the system that has been practiced in China for thousands of years. It is the system which is practiced now in the neighborhood of Nice, where they grow orange trees and scented flowers, and where they grow a large quantity of things which require rich manure ; and it is perfectly certain that it is the system in which there is least waste.

Now, this system has been very much revived of late in certain towns. In Edinburgh and Glasgow, and especially in many foreign towns — in Berlin, Leipsic, and in Paris — this is a plan which is adopted on a large scale. Of course, the difficulties of the system are enormous, and the nuisance is considerable. As far as the difficulties are concerned, I may tell you what Dr. Trench, of Liverpool, has calculated in regard to that town ; he calculates that the space that would be required for the spare receptacles for the borough of Liverpool would be 11 acres, 2 roods, 32½ perches ; that if put on a railway four-abreast they would extend a distance of 12 miles. Now, you see at once that a system which requires anything of that sort is not a system likely to be adopted for any large town. But, mind, there is no doubt whatever about this, that for small places it is an infinitely more healthy and more reasonable system in every way than either of the other two plans that we have considered. It is carried out with a simple bucket, in which the refuse matters cannot be allowed to remain for a long time, because it is not large enough, and because they would become too offensive. This system is evidently better for health than keeping the matters about the premises for a long time in any other form whatever. A variety of this plan is to be found in what are called the trough latrines that are used in many large manufactories. This simply consists of a trough which runs below a row of seats, which trough can be emptied into barrels by lifting a plug at the lower end ; the stuff is taken away to farms.

There are two or three varieties of the tub or pail plan. One is known as the Goux system, and another as the Eureka system ; these systems have been much praised of late. They are simply pail systems, in which some deodorizer or some absorbent is used. In the Goux system there is a sort of double pail, with an absorbent between the two pails ; and the idea is to do away with some of the offensiveness of the tub or pail system.

Now, I must say a few words to you about the dry earth system, which has been so much praised. In the first place I may tell you that sifted ashes are sometimes used instead of dry earth, because they are always at hand, and that there are several plans for the use of sifted ashes which are attempts to obviate the difficulties which are met with in the procuring of dry earth. It is found that when a sufficient quantity of dried and sifted earth, especially of particular kinds, is thrown upon refuse matters, that they are deodorized, and that they may be kept for a very long time without becoming offensive. The conditions are these : in the first place the earth must be dried, and in the second place it must be deposited on the refuse matters *in detail*, as it is called, that is to say, you must not take a great heap of excretal matters and throw a lot of earth upon it, but you must throw a little earth upon it each time the heap is increased by any more refuse matters. It is found, then, that about one pound and a half of dry earth is sufficient to deodorize the excretal matters that are passed at one time by an individual. With regard to the kind of earth, almost any earth will do except sand and chalk.

The next point is, that such earth may be used several times over. After it has been used once, it requires merely to be dried again and sifted, and you cannot tell it at sight at all after it has been used two or three times from that which has been used once ; you do not see any difference whatever ; all the organic matters and all the matters that would be offensive are entirely absorbed and rendered inoffensive to the smell, and so long as it is kept dry this earth remains quite inodorous.

There are all sorts of forms of closets, and so on, that have been contrived for utilizing dry earth in this way, but there is not the slightest necessity that I should describe these plans to you. I will therefore go on now to tell you of the results that have attended the application of this system at various places, and the advantages and disadvantages of the system. In the first place, with this system it is a *sine qua non* that no liquids are to be thrown into the earth closet, so that it is a system which does not provide for slops ; that is against it to begin with. Then if any liquids are accidentally thrown in, or if, as is the case in certain places, the air is exceedingly damp, or if the contents get moist in any way, you have, to all intents and purposes, a cesspool without its advantages, or without the special precautions that are commonly taken with regard to cesspools. That is another disadvantage of the system, we shall find more directly.

Now for the advantages. The advantages, perhaps, are best shown by giving some statements as to the working of the system at different places. At Broadmoor the lunatic asylum is supplied with earth closets. The water closets with which the place was originally supplied were done away with, and the earth closet system adopted. A mixture of earth and ashes is used, but the slops are allowed still to pass through the drains. Here, you see, you have got every advantage that such a system could have. You have sewers originally made by Mr. Menzies, made for the water closet system; and so they can send just what they like into them, and treat the earth closet in a sort of drawing-room fashion—if I may so call it—I mean give it its best chance.

Then at various schools it has been

found to answer very well. The simplest form of it is a mere trough into which the refuse matters fall and into which earth is thrown. At various jails the plan has also been used with considerable advantage. At one place, where there was an attempt made to save all the fæces and urine of the boys in the school in this way, it was found that four pounds of dry earth a day* was required for each boy. I mention that to you to show you the absolute impracticability of doing a thing of that sort on a large scale. The expense, of course, would be enormous.

Where the plan has been used as a temporary arrangement it has, on the whole, succeeded very well, and especially so, for instance, at Wimbledon Camp. At Wimbledon Camp there is no doubt it has been an enormous improvement on the old system. Now you see at once that that was a temporary

---

* That would be 125 tons a week for a population of 10,000.

arrangement, that they could get plenty of earth, and so there was very little to be wondered at that a system which does, if it is properly carried out, deodorize offensive matters should have been there so far a success.

Now, I must tell you a little about the Indian experience. The Indian experience has been unfavorable to this system, but there are many statements made in Indian reports to the effect that it is a considerable improvement on some of the systems that were in vogue before. That you would easily believe if I read you a description of some of the systems, or, rather, of the want of system, that they had before they adopted this plan. The Army Sanitary Commission make the following statement :—"It is insufficient to remove only one class or cause of impurities, and to leave the others ; and no sanitary proceeding which does not deal effectually with all of them can be considered as sufficient for health."

"The following sources of impurity require to be continually removed from inhabited buildings in India as elsewhere; (*a*) solid kitchen refuse including *débris* of food; (*b*) rain water which would if left in the subsoil tend to generate malaria; (*c*) all the water brought into the station except that which accidentally evaporates. This water is used for drinking, cooking, washing, baths and lavatories. The amount cannot be taken at less than twelve gallons per head for every healthy man, woman, and child, including servants; from thirty to thirty-five gallons per head for every sick man per day, exclusive of water for horses. . . . Practically, this water in all climates, but especially in India, becomes, if not safely disposed of, an inevitable source of disease and ill-health. It contains a large amount of putrescible matter, and if urine were mixed with it, it would become so noxious that it would matter very little whether or not the contents of latrines were added to this other sewage; (*d*) the matter from latrines, including solid and fluid excreta at about one pound per man per day, or, in round numbers, half a ton per day per thousand men."

Those are the matters which will require to be removed. Now the Commissioners go on to say that the solid *débris* being removed by hand or cart labor, the refuse water must "either be passed into cesspits, or it must be carried away, or it must be allowed to find an outlet where it can by surface drains—probably into the sub-soil."

Then, farther on, they say that the latrine matter, with which alone the dry earth system proposes to deal, "is to the fluid refuse of barracks, hospitals, cook houses, and so forth, as 1 to 190; that is, for every pound of human excreta removed under the dry earth system there are in every well regulated establishment about 190 of fluid refuse which must be otherwise disposed of." You see at once that this is absolutely condemnatory of the system for use in permanent barracks, and I think you will come to the conclusion that I have come to, namely, that it is a system that is only fit for temporary places, like the camp at Wimbledon. It is perfectly plain that it is absurd to have two systems, a system of sewers to carry away the foul water of a station—for that you must have—and another system for carrying away a certain portion of the excretal matter, which might all perfectly well be allowed to go away with the foul water.

As for the utilizing of it in this way, I do not believe in it at all. I mean to say that the cost of bringing in dry earth into a station, and especially into a large town, and then of carrying it away again, would be considerably greater than the money that would be got for the manure. You will see, and I dare say you have seen, that the manure collected under the dry earth system has been put down as worth all sorts of fabulous sums. Well, it is not worth anything of the kind. It is the greatest mistake to suppose so. The manure from the dry earth system is a good garden soil, and it is not anything more.

It is not a manure. And this earth that has been passed three times through the closets is nothing more than that, as you will see in the report of the British Association Sewage Committee. In that report there are given the results of analyses of the earth after passing once, and after passing twice, and after passing three times through the closet, and it is said that after passing three times through it is nothing more than a rich garden soil, and it will not pay for incurring the expense of carriage to a long distance; so that the utilization question is certainly not met by the dry earth system. This is what I wanted to come to.

I may as well tell you that it has been attempted to apply this system to a town. One part of Lancaster is supplied with dry earth closets, and there are several villages in which it has been tried. In villages it seems to answer very well when looked after. In a town, for some of the reasons that I have given you, it fails. There is no doubt about that.

That will finish our consideration of the systems which propose to separate the excretal refuse as if it were something totally, and entirely, and essentially distinct from all other refuse matter forming the sewage of a town. Those systems all go upon a wrong principle. This refuse matter is dangerous to health, and those system's, one and all, go upon the principle that these matters may be retained in and about houses as long as possible, so long as they do not create a nuisance, or so long as they are not felt to be a nuisance. Now that position is obviously wrong. All these systems depend upon leaving such matters as long as possible about the houses. The object of them all is to produce a certain result with as little expense as possible. It is perfectly plain that the longer this refuse matter is left about houses, under all these systems,—I do not care which one you take,—the cheaper the plan will be carried out, so that there is a tendency in all these systems to leave dangerous refuse matters about premises for a very long time. Now, the answer is that they are perfectly deodorized or disinfected, as the case may be. The answer to that is—if your system is perfect, they are deodorized in some cases; for instance, in the dry earth system they are deodorized. But if they are not, all the danger arises that could arise from any other of the bad systems I have described to you. Then, again, the fallacy is entertained that deodorization and disinfection mean the same thing. It is certain they do not. We know quite well that the dry earth system deodorizes refuse matters. They do not putrify and cause offensive smell after deodorization, but we do not at all know that that system disinfects matters; and there is not the slightest reason for supposing that this earth, if at any time rendered moist,—and we do not know whether or not disinfection takes place even when dry,—that this earth may not then be dangerous and have infecting properties. I mean to say we do not know that the excreta of cholera patients, or typhoid fever patients are disinfected, as well as deodorized, by this mixture with dry earth, and so I think you will all agree with me that the plan which has for its principle the removal of these excretal matters immediately from the vicinity of habitations (utilizing them afterwards, if possible), that the plan which goes upon the principle of removing them in the cheapest way possible, viz., by water carriage and by gravitation, removing them at the same time with all the other refuse matters of the population (with the single exception of the ashes), I think you will agree with me that that is, after all, the most reasonable plan. And I do not hesitate to say that nothing has contributed so much to lower the death rate of towns as the introduction of the water carriage system.

Before describing to you the composition of sewage and the ways in which it has been proposed to treat it, I have a few words to say to you about the construction of the apparatus in houses, especially as to the construction of the apparatus used for the removal of such effete matters from houses. I am going to say a word or two on this subject because of the importance of the points connected with it—points that every one of you ought to know.

In the first place, the simplest form of closet that can be used is one with an earthenware pan and syphon, all in one piece, which of course, so long as it is not broken, always retains a certain quantity of water in it. The advantage of this closet is that it can always be readily cleansed. In towns where a more complicated form of apparatus has been tried for the poorer classes and for persons who are not careful, water closets have always failed. One of the great arguments for the supporters of all the dry systems, and especially of the dry earth system, has been that the water system has failed because persons will not take reasonable care ; but that has been where the apparatus has been too complicated, as has often been the case in London. Now in towns supplied with apparatus of that sort there is much less risk of anything getting out of order ; anything that finds its way into the syphon can be easily got out again, and in fact nothing short of pushing an iron rod in, and making a hole in it, is likely to do any harm, and when a hole is made in it it is easily detected, because the water will not then remain in the syphon ; in fact nothing is easier than to discover such a damage.

In the next place they are cheap. One very important point about this system, especially if these closets have to be inside houses or dwellings, is that there should be a hole in the syphon, at the highest point of the pipe above the water, and leading into the drain, and that at this point there should be attached a ventilating pipe. Any sewer gases arising will then be taken off by that pipe, which should be carried to a sufficient height and turned over at the end.

As to the water supply any simple apparatus will do for that ; the usual plan is to have a wire, which, when pulled, lifts a plug in the cistern, and water runs down a pipe which generally ends in the side of the pan, the aperture being so directed that it whirls the water round the pan ; and the waste pipe of the cistern supplying these may, if that cistern is outside the house, and is only used to supply the closet, be made to end in the same supply pipe. There is very little harm in that, but it should not be done if the same cistern is used for supplying drinking water, which ought not to be the case, although it so often is. A more complicated form of water closet, which is commonly used, requires a word of notice. In this sort you have what is called a D trap, and above that there is what is known as a container, which is a large iron vessel opening below into the water in the trap. Water always remains in this D trap up to the level of the outlet. It is called a D trap from its shape ; it is like a D placed thus The pipe which leads from the container (which is the iron vessel immediately under the pan, and in which the basin moves) dips under the surface of the water in the D trap.

Now a few points of caution about this method are necessary. This is the apparatus which is accused of having brought us a large amount of typhoid fever, diarrhœa, and even cholera in large towns, which we should not have had otherwise, and no doubt to a certain extent the system is to blame for it. And I am going to show you where it is to blame for it, and what precautions we have to take to prevent this.

One way in which it is to blame is that the descent pipe, called the soil pipe, is a very convenient place to make the waste pipe of the drinking water cistern end in, and so, very frequently in houses this waste pipe comes down and ends there. The soil pipe goes out of the D trap, and then joins the main soil pipe of the house, or becomes itself the main or perhaps the only soil pipe. This main soil pipe is very seldom open at the top, unless it has been purposely so constructed, and so any foul air in the drain below, or in the soil pipe itself cannot get away, and so it simply goes up the waste pipe of the drinking water cistern, and in fact the waste pipe of that cistern forms the ventilator of the soil pipe, and any poisonous matters in the air in it are absorbed by the water, and drunk, and this is unquestionably one of the causes of the spread of typhoid fever. The waste pipe, however, should not end there, but should end, as I have before told you, outside in the open air.

Now supposing there is no waste pipe ending there, the foul air which accumulates in the soil pipe will have a good many of its ingredients absorbed by this water in the D trap, and they will be given out at the surface of the water into the "container." As soon as the apparatus is worked, and the pan let down so that the water runs out of it, then the foul gases, which have been collecting under pressure in the container, immediately issue into the house.

Now how can that be prevented?

There are two ways of preventing it. In the first place, this soil pipe should be open at the top, and then you will never have sewer air with pressure on the D trap. That is quite clear. Or, if you do not carry the soil pipe itself up to the top, there should be, say, a 1-inch leaden pipe going from it up to the top of the house, and turned over, ending at some convenient place, not near the outlet of the chimney. So thus you prevent any foul air from collecting in the soil pipe and rendering the water in the D trap fouler than it need be; but the D trap is always full of water, the apparatus is seldom worked so long as to replace all that water, and so it remains always more or less foul. The foul matters in the D trap putrefy, and so foul air collects in the container, and as soon as the apparatus is worked, this foul air in the container immediately rushes out, because it has been collecting in considerable quantities, a thing which you must all have observed over and over again. That can be prevented perfectly well, so well indeed, as to render closets manageable even in the most inconvenient situations in which they can be placed, and in such situations they frequently are placed in many of the large houses in London; underneath staircases, and close to drawing rooms, or even opening directly out of bed-rooms, and in such places. Although they should, if possible, be removed from such situations, they can be made perfectly sweet in a very simple way, and that is done by making a hole in the container, attaching a small ventilating pipe to it, which pipe is taken through the wall of the house and made to end in some convenient situation, and then you never get any foul air accumulating under pressure, to rush out when the pan is moved and the water in it let fall into the container.

The valve closet is a great improvement upon this, inasmuch as the container is merely a small box in which the valve works, whereas the volume of water used is much greater.

Those are the chief points about this

rather complicated apparatus, which is evidently not fit for the use of careless persons. Now for the kind of apparatus fit to be employed when large numbers of persons use the same place. There has been an apparatus contrived called "The Trough Water Closet." I spoke to you before about "Trough Latrines." They are not water closets; they are constructed in very much the same way, but in the "Trough Latrines," the excreta are collected for the day, and then are emptied into a cart and taken away. That is a modification of the "Pail System."

The "trough water closet" may be briefly described thus. Underneath the row of seats there is a trough made of iron or slate, or any convenient material of the sort. This trough at its lower end has a connection with a sewer, the mouth of which is fitted with a plug; which plug can be moved up and down by a lifting apparatus in a separate compartment, which can only be got at by the person who has charge of the place;

because, of course, among large bodies of men, there must always be a man appointed to have charge of this, just as in the case of the earth closets. This compartment can only be got at by this particular man, and he has access in order that he may lift up or let down the plug.

At the other end you have a water tap supplied from a cistern ; the man who has charge of the place comes at night, lifts up the plug, lets the contents all run away into the drain, then washes out the trough, lets down the plug, charges the trough with a little water, and leaves it till the next day. That is the "Trough Water Closet," and that is the most convenient form of closet for use by large bodies of persons, especially of careless persons.

As an instance of the success of these closets, I may mention the town of Liverpool. Dr. Buchanan and Mr. Radcliffe, say—" Nothing could be more admirable than the working of the Liverpool arrangement, and nothing could be more marked than the difference between them and what are called water closets, in the poor neighborhoods of London and other large towns." Dr. Hewlett also gives a favorable opinion with regard to these closets. He says—"The trough water closets in use at Liverpool, and the self-flushing tumbler water closets at Leeds, where they answer remarkably well, appear to me to be the best kind for use in poorer districts, especially for closets which are frequented by more than one family." These opinons are sufficient.

The tumbler water closet is very nearly on the same principle. There is a very nearly level trough with a connection with the drain at the lower end ; at the other end there is a sort of swing bucket which is placed below a tap. The water is running from this tap continually, but slowly, and the rate at which it shall run is subject to arrangement. As soon as the bucket contains a certain amount of water, it tips over, empties its contents into the trough, washing away whatever is in the trough down into the drain. This plan is also reported to be an excellent one. Of course the water supply and the buckets are placed in a separate compartment, and can only be got at by one person.

We pass on, now, to consider the composition of sewage. This has been well stated in the first report of the Rivers' Pollution Commissioners, in the following words "Sewage is a very complex liquid ; a large proportion of its most offensive matters, is, of course, human excrement, discharged from water closets and privies, and also urine thrown down gulley holes. Mixed with this, there is the water from kitchens, containing vegetable, animal, and other refuse, and that from wash-houses, containing soap, and the animal matters from soiled linen. There is also the drainage from stables and cowhouses, and that from slaughter houses, containing animal and vegetable offal. In cases where privies and cesspools are used instead of water closets, or these are not connected with the sewers, there is still a large proportion of human refuse, in the form of chamber slops and urine. In fact sewage cannot be looked upon as composed solely of human excrement diluted with water, but as water polluted with a vast variety of matters, some held in suspension, some in solution."

Now there are great variations in the composition of sewage at different times of the year, and also at different times of the day and night. But there is not a great amount of difference between the composition of the sewage of towns where there are water-closets, and the composition of the sewage where there are not. In the first report of the Rivers' Pollution Commissioners, it is shown that there is "a remarkable similarity of composition between the sewage of midden towns and that of water closet towns. The proportion of putrescible organic matter in solution in the former, is but slightly less than in the latter, whilst the organic matter in suspension is somewhat greater in midden than in water closet sewage."

I must now give you an account of what the average sewage may be taken to be. You may take it that an average sewage has this composition ; in 100,000 parts of it there are about 72 of total solid matters in solution, which total solid matters include between four and five of organic carbon, something over two of organic nitrogen, from six to seven of ammonia, and from ten to eleven of chlorine. Besides these 72 parts of dissolved matters, it contains 44 or 45 parts of suspended matters, of which about 24 are mineral, and 20 or 21 organic. Now that is the average. There are extremes. The variation of the London sewage in total combined nitrogen, is from three parts to eleven in 100,000, so that you see there are very considerable differences. And this is partly due to the fact, which is plain enough, that there is a greater amount of refuse thrown into the sewers at one time than at another, but still more to the great variations in the amount of water. As an instance of variation with the time of the year, I may tell you that during the winter before last, the average amount of ammonia in 100,-000 parts of the sewage of Romford, was from five to six parts, whereas in the previous summer, the average was only two and a half to four. So that the value varies considerably at different times of the year. It varies because a considerable amount of rainfall is allowed to get into the sewers. It would vary little in towns where very little rainfall is allowed to get into the sewers, and where the water supply is pretty constant throughout the year. The variation in composition during the day and night is very important, and during the night, in many towns, the sewage is very little more than water.

Now as to the value—you may calculate the value of sewage in two ways. Thus you may calculate it approximately, from the number of persons who contribute to make it, and from the value that we have assigned to the refuse matters coming from each person during the year. We have assigned 8s. 4d. a year for these matters, but we will take the lowest value ever assigned to them, which is, that the annual excreta of a human being, taking an average of all ages, are worth 6s. 8d. a head. That value was assigned by Messrs. Lawes and Gilbert, and they have never been valued at less. So that, taking no other refuse at all, if you can get at the excretal refuse matters of à population of three millions, it ought to be worth £1,000,000 per annum, as far as that calculation goes.

But you may calculate the value, again, from the composition of sewage itself. And if you do that, you will find that the money value of the substances dissolved, say in 100 tons of average sewage, is about 15s., while the money value of the suspended matters is only about 2s. The value, then, of these constituents, is about 15s. for the dissolved matters, and 2s. for the suspended matters ; that is to say, that 100 tons of average sewage are worth 17s., or about two pence a ton.

If you consider that you do not always get average sewage, or sewage of an average composition, and that very often the sewage is extremely diluted so that, instead of there being something like the dry weather average of sixty tons per head per annum, you often get 100, or even more, it is plain that we must not take so high a value as that I have just stated for it, and so it is usual to take a value of one penny per ton, instead of two pence. If then you take this sewage at a penny per ton, as containing on an average about four grains of ammonia in a gallon, as it does, or between five and six in 100,000 parts, as I said before, then you may consider that sewage is worth one farthing per ton for every grain of ammonia per gallon it contains.

And again, if you take the total amount of sewage of three millions of persons at an average dilution of about 80 tons per head per annum, and put it at the value of 1d. per ton, you will find it comes to almost exactly the same as the calculation made the other way, viz., something over £1,000,000.

Now what are we to do with this sewage which has the value which I have just assigned to it, that is to say, which has that value if you can get the manurial properties out of it? The general plan at present is to turn it into the rivers. This plan has arisen because the sewers we use were originally built for drains, and meant for drains, and therefore naturally discharged into the rivers, and it is no doubt from this circumstance that we have so many attempts to keep a certain proportion of the manurial refuse out of the sewers by means of midden closets, and pail closets, and earth closets and so on, and also with a view to the prevention of the fouling of the rivers. Well there are two kinds of evils that arise from the fouling of rivers, two especially, but there are plenty of others in addition. The first is that these rivers, even when they are large ones, get to a certain extent blocked up by the sediment that is deposited from the sewers; this is the case even with the Thames. In the year 1867 it was pointed out that there was going on a formation of extensive shoals in the River Thames outside the main drainage outfalls near to Barking Creek and Crossness. These deposits were very extensive. Near the southern outfalls for instance a depth of fully seven feet of deposit was found, and in fact it was going on to such an extent that it threatened to interfere seriously with the navigation. Plans can be seen which show that a narrowing of the bed of the stream had been going on even since this was pointed out in 1867. Besides that, it was also shown that the tide did not carry away the matters suspended in sewage. Experiments were made by Mr. Frank Foster first, and afterwards repeated by Mr. Bazalgette and Captain Burstal—which show that suspended matters, floating bodies, were carried down by the tide to a certain point, and then carried up again farther than the point at which they were originally thrown into the stream, and it is a fact that a certain amount of sewage deposit takes place above the outfalls into the River Thames.. Now in small streams as well as in navigable rivers this is of course a very serious matter. That is the first thing. Then perhaps a less important matter, but still one of some importance, is that fish are killed in rivers into which sewage is turned. They are not killed by fresh sewage, but they are killed by the gases which are given off by the decomposing deposit at the bottom of the river, by sulphuretted hydrogen especially.

Then the next danger is the pollution of the drinking water of towns lower down on the rivers, and this has gone on to a very considerable extent, to such an extent that at last Londoners have found out what they are drinking, and all the towns on the River Thames have got injunctions to prevent them turning their sewage into the river. This comes to a climax when you have a case where a town actually turns its own sewage into a river at a particular place, and a mile further down takes out its water supply. That occurs in a town in England at the present moment.

If the towns are not to turn their sewage into a river, what can they do? You see the sewage contains suspended matters and dissolved matters, and both among the suspended matters and dissolved matters are substances that are injurious to health if drunk with water. You see also that the dissolved matters are considerably more valuable than the suspended matters, in the proportion of 15 to 2. But this was not always known, and so the first attempts at purifying the sewage consisted of simply straining it. The sewage was strained and the suspended matters were thus separated and were then sold as manure, or mixed with town ashes and sold for manure, and the somewhat clarified sewage was then allowed to escape into the stream. That is the practice carried on in a great many towns at the present time. The suspended matters are worth comparatively little, and you lose the best portion of the manurial matters. In the second place the purification of the stream is only partially effected, because the clarified sewage that runs into the stream putrifies after it gets there, and you get the stream fouled to a very considerable extent, so that that plan is evidently not sufficient.

Then come different chemical processes. The purification was attempted by various chemical processes, and it is still attempted to precipitate the valuable ingredients dissolved in the sewage as well as the suspended matters. Now there are plenty of ways in which you can clarify foul water, but you see at once that it is not so easy to precipitate those particular matters that are in solution in sewage, because you see in the first place that the most important constituent, or at any rate one of the most important constituents, the most important from its quantity at any rate, is the ammonia. You know perfectly well that you cannot precipitate salts of ammonia on a large scale at all from a dilute solution, and you will see, therefore, at once, that all attempts to precipitate the valuable matter of sewage are likely to fail, even from that cause alone. Then in the next place you have organic matter in solution. Now we do not know of any substance at present which can be used on a large scale at any rate, that can be relied upon to precipitate organic matters in solution, especially organic matters in the state in which they are in sewage, viz.: in a state of very rapid decomposition, and these are the substances matter which are most dangerous, and which have to be separated, so that you will be prepared to find that most of the precipitation processes have failed. You will find a long description, and an excellent one, of most of these processes in the Second Report of the Sewage Commissioners, published in 1861, giving the results of many analyses. Several of these processes are capable of precipitating at any rate one important ingredient in sewage, and that is the phosphoric acid, an important ingredient which can be precipitated in several ways, and they also—some of them—precipitate some of the organic matters.

Now these are some of the more important precipitation processes brought before the public. In the first place there is the lime process which was practised at Tottenham and Leicester, and some other places, and which merely consisted in adding a certain proportion of milk of lime to the sewage. The result of this process was that no element of agricultural value that was in solution was precipitated by it except the phosphoric acid. The suspended matters were very fairly well removed from the sewage (that you can do perfectly well by straining), and sometimes the amount of organic matter in solution was increased, because some of the organic matter originally in suspension passed into a state of solution, which it always will do by mere agitation, and also the amount of ammonia contained in the sewage was increased, so that by that process, as well as by some others, the water discharged into the river sometimes contained actually more impure ingredients than the sewage contained in solution, some of the organic matters in suspension having passed into a state of solution. A fault of the lime process is that the precipitated matter remaining is alkaline, so that much of the ammonia it contains is given off, and the next thing is that it is nearly worthless.

The Rivers' Commissioners pronounce it "a conspicuous failure, whether as regards the manufacture of valuable manure or the purification of the offensive liquid." The next that I have to mention is a variety of the lime process, in which lime and per-salts of iron were mixed and used. This is a much better plan, because the per-salts of iron will fix the sulphuretted hydrogen and all the phosphoric acid. The fault of this plan is, that it does not precipitate anything else that the lime process did not, and its virtue is this, that it deodorizes the liquid and the precipitate. Salts of iron have been used alone, and they do without doubt deodorize the water, and precipitate the phosphates and the suspended matters, but they only delay the decomposition of it, and again they are too expensive.

Then several processes in which clay was a precipitating ingredient may be mentioned. In the first place, Holden's, in which sulphate of iron, lime, and coal dust, with some clay are used, and Anderson's, which is very much the same as Bird's, which consists in the addition to the sewage of crude sulphate of alumina. Stothert's consists of the addition of sulphate of alumina with sulphate of zinc and charcoal.

And, lastly, the celebrated A. B. C. process. The A. B. C. process was so called from the chief ingredients that were used, with the object of precipitating the sewage, namely, alum, blood and charcoal. You have all probably heard sufficiently about the A. B. C. process lately. You know the Company has attempted to purify some of the sewage of London, at Crossness, and no doubt you have heard that a combined report has been issued by the Engineer and the Chemist of the Metropolitan Board of Works, which report shows perfectly well that although the sewage was at any rate clarified, and although there was a certain amount of purification effected (we can't say exactly what amount, as in this report we have not the analysis of the original sewage); although that was the case, the manure produced was not worth more than twenty shillings a ton, while the cost of producing it was £6 6s. 4d. ! Then there is a process known as Hille's process, which is chiefly a deodorizing process. A mixture of lime and tar, and chloride of magnesium is used ; the precipitate is of very little value. Carbolates and sulphites of lime, and magnesia, have also been proposed as precipitants which would also deodorize the sewage. At Carlisle, carbolic acid is used to deodorize the sewage.

There are two or three processes in which phosphates have been used. The idea of using phosphates to precipitate sewage was this,—that the precipitate produced by other substances, like lime and clay, which are useless as manures, will not sell, because it will not bear the cost of carriage; but if you add a substance which is itself a manure, and precipitate the suspended matters with it, then they would sell, and then you would get a manure that is worth carrying.

Now, the first phosphate process has been proposed over and over again. In England it goes by the name of Blyth's process, and the principle of it was this; —there is a salt of phosphoric acid (to wit, the phosphate of magnesium, ammonium and hydrogen, a triple phosphate), which salt is insoluble in water containing salts of ammonia, and it was thought that by adding a salt of magnesia and super-phosphate of lime, or super-phosphate of magnesia and lime water, to sewage, that a precipitate of this triple phosphate would take place. The result was that it was found to be the most expensive process ever adopted, and that a great proportion of the phosphate added went away in the effluent water. The salt in question is not at all insoluble in pure water. It is only insoluble in water containing an excess of ammonia; so that the condition for the success of this experiment was that the water turned into the river was rich in ammonia — an obvious condition for failure of the experiment—and the result was the loss of a great amount of the substances added. That process has failed over and over again.

Then we have a phosphate process patented by Messrs. Forbes and Price. In this process an insoluble phosphate of alumina in large quantities is used, and it is rendered soluble by being mixed with strong hydro-chloric acid. This is mixed with sewage, lime water then is added, and the result is that the suspended matters are carried down very completely, and the sewage is left very clear. All offensiveness is entirely taken away; the effluent water passes off containing all the ammonia that the sewage contained before, and at any rate the greater portion of the organic matter in solution. This process, therefore, could only be used as a preliminary process to some other treatment. I will not say any more about that.

Then recently another phosphate process has come forward, called Whitthread's. That process has been reported on by the Committee of the British Association appointed for the consideration of the treatment and utilization of sewage, and that process is the only precipitating process with respect to which it has ever been said that it does precipitate most of the organic matter that is in solution. It precipitates all the suspended matters, and so far as the preliminary experiments, which were carried on under the supervision of the Committee of the British Association,—as far as those preliminary experiments go, this process depends upon the use of a substance known as di-calcic phosphate, a particular form of phosphate of lime, which seems to have the property of carrying down organic matters in solution. The deodorization is also complete. It does not in any way remove the ammonia in solution, and it remains to be seen whether that process, or indeed any other process, is capable on the large scale of so removing the organic matters in solution that the liquid may at any rate be harmless after it is thrown away.

Lastly, I have to mention to you General Scott's process. General Scott's process consists in mixing the sewage with a certain amount of lime and of clay. It has been reported on by the Rivers' Pollution Commissioners and by the British Association Sewage Committee. About 10 cwt. of lime and 8 cwt. of clay are added to 400,000 gallons of sewage. This mixture of lime and clay is added in considerably greater proportions than the precipitants are added under the other processes. With the others you add as little as possible. With General Scott's process you add a great deal. Well, this mixture is added to the sewage in the sewers before it gets to the tanks, and the result is that the sewage is entirely deodorized, and as soon as it arrives at the precipitating tanks and is allowed to settle, the whole of the suspended matters, including the lime and clay which have been added, are deposited at the bottom of the tanks. This deposit is run out in a semi-liquid condition as soon as there is enough of it. It is then dried, or it may be compressed by what is known as Needham and Kite's Press. Needham and Kite's Press is a press which has a number of canvas bags in it, into which bags this mud is run. They are then pressed together by a hydraulic press. A certain portion of water is thus squeezed out of the mud, leaving it in a comparatively dry state. It is then taken up in lumps, dried by heat if necessary, and placed in a kiln. A fire is lighted below it with a small quantity of coal, and it burns.

"The area is laid out in square beds intersected with roads and paths, along which are constructed the main carriers which receive the sewage from the outfall sewer and distribute it over the beds." As soon as it is once set alight there is no necessity to put any more coals in the kiln. The sewage deposit with the clay and lime is supplied from the top of the kiln, and it is gradually taken out as it is burnt, through an opening in the bottom, and no more coal is required. There is sufficient organic matter in the deposit for it to go on burning, when once well lighted, for any length of time. The result is the production of a cement, and an excellent cement. This cement can be made of different qualities, and it certainly answers perfectly well as a cement, and the process causes no offence.

The result on the sewage is that it is clarified, and the phosphoric acid contained in solution is precipitated, so that this cement contains phosphoric acid. The ammoniacal salts are left in solution, and the organic matters in solution are not touched by the process, or rather they may occasionally be increased from some of those matters in suspension passing into a state of solution. The cement prepared can be used as ordinary cement ; or it has been suggested by General Scott that in places where lime is already used as a manure, it would be considerably better to use this sewage lime, after it has been calcined, on account of the proportion of phosphoric acid in it. It is a process, then, that does not at all pretend to purify the sewage; it merely pretends to afford a means of dealing with the suspended matters of the sewage, and to leave it in a condition in which it is better fitted for treatment afterwards.

We pass on now to the remaining methods for the treatment of sewage, which depend upon the filtration of it through soil. I have described to you the effects of the filtration of foul water through gravel and sand and charcoal. You may say this can be done for water containing a small amount of impurity, but can it be done for water containing a large amount of organic matters both in suspension and solution, as is the case with sewage? Now, filtration may be of two kinds; at any rate there are two principal kinds, downward and upward. You may either pour the water on to the surface of the filter and allow it to pass through, or you may conduct the water underneath the filter, and let it rise up through the filtering material. By the first process, which is known as downward filtration, sewage can be satisfactorily purified on one condition, namely, that the filtration shall be intermittent. I told you how a filter purifies, and you will see at once that this is a necessary condition for the purification of sewage. If you have sewage falling on a filter bed and passing through it to drains below, the organic matters in that sewage are only oxydized, if there is air in the filter; and there cannot be air in the filter unless your process is intermittent. But if your process is intermittent, when you stop pouring sewage on to the filter bed, the remaining water trickles down into the drain below, and so fills your filter with air. You must have an intermittent process.

That shows you again why upward filtration is not capable of purifying sewage. Supposing you have water admitted underneath the filter, and that it is so constructed that it can rise up to the surface of the bed and flow off it, your filter bed is always charged with water. Upward filtration then does not afford the means of aërating the water. By intermittent downward filtration we have a means—as pointed out by the Rivers' Pollution Commissioners (1868) —a means of satisfactorily purifying sewage.

Experiments conducted by filtering London sewage through 15 feet of sand showed in the first place that " the process of *upward* filtration through sand is insufficient in the purification of sewage from soluble offensive matters ; . . . . . . . . on no occasion was the effluent water in a condition fit to be admitted into running streams," but that the " process of intermittent *downward* filtration through either sand or a mixture of chalk and sand effects a very satisfactory purification of sewage when the sewage treated amounts to 5.6 gallons per cubic yard of filtering material in 24 hours ; but that the purification becomes uncertain and unsatisfactory when the rate of filtration is doubled, that is when the sewage treated amounts to 11.2 gallons per cubic yard in 24 hours." And so on. And then the amount of purification is given, and the value of different soils as purifying agents. Now, filtration through charcoal has been used in one process, the process known as Messrs. Weare's process ; the process has been used in the filtration of the sewage of the workhouse at Stoke-upon-Trent. One would have expected, if the experiment had been conducted properly, that it would have been a success. However, all the sewage of that particular place was exceedingly strong, and although it was purified to a considerable extent by Weare's process, it is not reported favorably on in the report of the British

Association Committee, and it has not been employed on a larger scale. Then intermittent downward filtration through soil has been employed as a means of purifying sewage on a very large scale at Merthyr Tydfil by Mr. Bailey Denton, and this has been reported on by the Rivers' Pollution Commissioners and also by the British Association Committee.

"Merthyr-Tydfil contains a population of 50,000—I am now coating from the proof sheets of last year's report of the British Association Committee—but according to information supplied to the Committee, the excretal refuse of not more than two-fifths of this number is discharged into the sewers, although the slops and other liquid refuse from a further like number (20,000) is stated to be admitted. It is not surprising, therefore, that the sewage is, as afterwards appears, weak." "An area of about 20 acres has, under the supervision of a member of the Committee, been converted into a filter bed for the practice of the system of downward filtration originated by the Rivers' Pollution Commissioners, as above described." "The soil of this area consists of a deep bed of gravel (probably the former bed of the River Taff, which is embanked up on the east side and is raised above the valley) composed of rounded pebbles of the Old Red Sandstone and Coal-measure formations, interspersed with some loam and beds of sand, forming an extremely porous deposit, and having a vegetable mould on the surface."

"The land has been pipe-drained at a depth of less than 7 feet, and the pipes are concentrated at the lowest corner, where the effluent water is discharged into the open drain which leads to the river Taff at some distance down the valley."

"The sewage before entering the farm is screened through a bed of 'slag' which arrests the coarser matters. It is applied to the land intermittently, for the area being divided into 4 plots or beds, it is turned on each one for 6 hours at a time, leaving an interval of 18 hours for rest and aëration of the soil." So that you see the right principal is carried out there. When the Rivers' Pollution Commissioners reported on intermittent downward filtration through soil, they said that it could be used to purify sewage. They also said, and it was so thought, that the sewage would be entirely wasted ; that the greater amount of it is wasted, as you will directly see ; but that it need not be entirely wasted we can see from these experiments at Merthyr-Tydfil, where large crops are grown upon the limited area.

"The surface of the land was cultivated to a depth from 16 to 18 inches, and laid up in ridges, in order that the sewage might run down the furrows, while the ridges were planted with cabbages and other vegetables."

Well now, this process has been carried out on a large scale, and it has been examined with the following results by the "Rivers' Pollution Commissioners," and also by the "British

Association Committee," and both sets of examiners have come to the conclusion that the purification is satisfactory. I am not going to give you the numbers, but I am just going to give you one or two facts about it. In the first place, when you look at the analysis of the sewage of a filter bed like this, you always have to take into consideration the possibility of dilution of the sewage with subsoil water, and in this place the dilution of the sewage with water from the river Taff is considerable. In the summer it was diluted with certainly more than an equal volume of subsoil water, and the gaugings in the winter showed that each gallon of sewage had become mixed with about 2 gallons of subsoil water. When this is allowed for, if you compare the results of the analysis of the effluent water with that of the analysis of the sewage, you find first that the suspended matters are all removed. Then with regard to the dissolved matters, the nitrogen, instead of appearing as it does in sewage as ammonia and organic nitrogen, appears as nitric acid ; it has very nearly all been oxydized,—a result that we get from purification of drinking waters by filtration ; but the importance and interest of the matter is, that after making allowances for dilution with subsoil water, the total amount of nitrogen in the effluent water is almost exactly the same as the total amount of nitrogen in the dissolved matters in the sewage, although in a different condition ; that is to say, that the nitogen retained by the land is almost exactly equivalent to the amount of nitrogen in the suspended matters. The effluent water, I may tell you, was so pure, both in the winter and in the summer, that in the winter nearly all the nitrogen in it was in the form of nitrates and nitrites, and in the summer $\frac{4}{5}$ths of it was in the same oxydized and harmless condition.

We have now to consider the subject of sewage irrigation. I have shown you that by filtration through the soil in a particular manner, sewage could be satisfactorily purified ; could be purified, in fact, so that the water which had passed through a filter of sand, gravel, or soil, was, practically speaking, drinking water. It is perfectly plain, therefore, that if you enlarge the area of your filter, and pass the sewage through a certain depth of soil, you can in that way, even without the action of plants, satisfactorily purify sewage. But as irrigation farms existed before intermittent downward filtration was thought of, it is neccessary for us to consider whether it is sufficient merely to turn the sewage on to unprepared land—whether it is sufficient that this should be done without making it a necessity that the sewage should pass through the soil.

There are now, at any rate, two classes of irrigationists. One set tells you that an irrigation farm is nothing in the world but a very large filter ; that it is absolutely necessary for the purification of the sewage at all times of the year that the sewage should pass through the land. I mean to say they will tell you that at certain times of the year, at any rate, if the sewage does not pass through the land, it will not be satisfactorily purified, and that there is danger of its not being satisfactorily purified at any time.

Others again say that it is not necessary that the sewage should pass through the soil into drains, and that it is not even necessary that the land should be drained in many cases at any rate. In the first place, there can be no doubt that upon almost all impervious soils, sewage can be purified by surface action to a very large extent indeed. At some of our sewage farms they work upon this principal. The sewage does not pass through the soil at all. On the soil plants are growing, and they take up the organic matters, ammonia, &c., from the sewage ; and the water which passes off, the overflow, is remarkably pure. But it will remain for us to consider bye and bye, whether, when plant action is least, this would be the case, whether in such cases the effluent water should not be left in an impure condition.

Well, now you have sewage brought on to a farm, if you are able to do it, by gravitation, if not, by pumping. On the farm tanks are constructed; as a rule, two tanks, the one being merely to be used while the other is cleaned out. Tanks are considered by many persons as not specially necessary, but they are so, both for the separation and collection of the grosser suspended matters and also for storing the night sewage. The sewage is run out from one of the tanks neither from the top nor from the bottom. At the bottom the sediment is allowed to deposit. At the top the scum accumulates, covers over the surface of the liquid, and to a great extent diminishes the offensive ordor. The sewage is allowed to run out between the scum at the top and the sediment at the bottom. One of the simplest ways of effecting this is by means of a kind of floodgate (such as one you may see at Brenton's farm, near Romford), made of pieces of board slid down one over an other; rings are fixed to the sides of these boards, so that one or more of them may at any time be lifted a little by means of a rod with hooks at the end, and then the sewage will run out through the gap made ; the lower part of the flood gate keeping back the sludge, and the upper boards keeping back the scum. The sewage flows either directly into the *carriers*, or when it has to be pumped, into the pumping well.

To take the sewage on to the land from the tanks, you may use concrete carriers, as they are the easiest made, and the cheapest. If you want the work to look particularly well, you can use brick-work, or earthenware, at the Tunbridge Wells.

If the carriers have to be lifted above the ground, as where there is a pumping station on the farm itself, they are best made of sheet iron, supported on wooden tressels. They must have simple taps which can be opened by merely taking out a plug.

These carriers run in directions which depend upon the slope of the land.

In the first place, I may tell you that the best sort of land to irrigate is flat land—quite flat—and then that which gently slopes. The main carriers are to run at right angles to the slopes of the land. They are carried under the roads by means of inverted syphons, and then the land is divided at right angles to these carriers into parallel beds.

I am now describing to you the plan which I believe to be the best.

The land is arranged in ridges and furrows, the crests of the ridges running at right angles of the main carriers and down the middle of each bed, so that each bed slopes slightly from the middle towards each side. At Brenton's Farm, where you can see the plan at work, the beds are 30 feet across. Along the top of the ridge there is run a minor carrier. This is merely a groove made along the crest of the ridge by a plough. The taps on the main carriers are just opposite to the beginning of these minor carriers, and the sewage can be let out of the taps and allowed to run along the minor carriers. When the minor carrier is full the sewage overflows and runs over the bed down each side into the furrows between it and the adjacent bed. It may be allowed to run into the minor carrier as long as no pounding occurs in the furrow. That is the "ridge and furrow plan."

There is another system called the "catch-water system." In that plan the sewage is taken in carriers along contour lines. The carrier along the high contour line is filled, and the sewage stopped at a particular place; it overflows and runs down the slope of the land into the carrier below. You can see that at work at many irrigation farms.

There is a variety of that called the "pane and gutter" system. I do not know that I need explain it in detail. The land is divided into pieces or "panes," running down the slope, and at right angles to the main carriers, and the sewage is run over the surface of these "panes" from the higher carriers into the lower ones.

There has been for a long time at Milan a plan of simply flooding the whole of the land, but in this way marshes are produced. The first obvious disadvantage of the catch-water and pane and gutter plans is that some land gets much more than the rest; because all the sewage that flows over the lowest level of a bed must pass over the whole bed from the top. If the land below gets enough, the land above gets too much, besides the fact that the lower beds get all the water that flows off the upper ones. On the other hand, with the "ridge-and-furrow" system I described before, you can just allow each particular bit exactly the amount it wants.

A boy goes along the carrier, turning on and off the taps as they are wanted, and a man walks up and down the ridges and stops the sewage at intervals, so that it overflows the minor carrier on each side and runs over the bed.

Any channels that convey sewage may be open. There is no reason whatever for covering them over. The loss of ammonia, &c., from the sewage is perfectly inappreciable even after a passage for a very long time through the open air, as proved by the experiments of the Rivers' Pollution Commissioners. You know already that by passing it through soil sewage can be purified. Now, a few words about passing it over the surface of the soil. The Committee of the British Association have made experiments upon this very point ; and I have comparisons of the results of the purification of sewage during very severe winter weather at some different farms. Now the first of these is Beddington Farm, Croydon. Here the analysis showed "that the nitrogen that is lost on this farm is lost for the most part in the form in which it came into the land, and that mere surface action (which is relied on here), is not sufficient to cause the oxydization of the ammonia and organic matters contained in the sewage. At the same time the purification effected was certainly very considerable."

Then again, the effluent water at the Norwood farm during the severe frost, was, practically speaking, sewage containing nearly half the amount of the ammonia that the sewage put on the farm did. It contained very little nitrogen in the form of nitrates. You know I told you that the main action of a filter was the conversion of ammonia and the nitrogen contained in organic matters into nitric acid. So that the amount of nitric acid in effluent water is a test of the oxydizing action of the filter. At the same time the analysis of the effluent water at Brenton's Farm, Romford, showed that the purification was very satisfactory indeed, for the effluent water only contained a very small quantity of actual ammonia, that is to say, about 0.14 parts in 100,000, as against 5.6 contained in the sewage, and of albuminoid ammonia only 0.059 parts remained out of 0.524 in the sewage, while the effluent water contained no less than 1.2 parts of nitrogen in the form of nitrates and nitrites.

In winter, when little action of vegetation is going on, mere passage *over* the soil will not purifiy sewage satisfactorily. The effluent water which goes off the land, is, to all intents and purposes sewage. On the other hand, there is a perfectly clear proof, that during winter if you pass the sewage *through* the soil, as at Brenton's Farm, purification goes on just the same as at any other time of the year when vegetation is growing.

Even during the summer there is a risk that the sewage may not be satisfactorily purified where the catch-water principal is adopted. This happened at Reigate, where the sewage was passed over one field and then over another, and the effluent water that came off the lower part of the second field was actualy more impure in several ways than that which came on to it from the first field ; that is to say, that this second field was to a certain depth so absolutely saturated with sewage, that the water, or sewage, or whatever you like to call it that came from the first field actually became more impure the further it went. It was in fact made stronger by the ,amount of evaporation which went on, and is another conclusive proof that surface action is not sufficient.

These results seem to me decisive in favor of the construction of irrigation farms as large filters. Then we come to the practical point—that it is therefore necessary that they should be drained.

On this head the British Association Committee express the following opinion :—

"It may seem almost superfluous for the Committee after so many years of general experience throughout the country, to argue in favor of the subsoil drainage of naturally heavy or naturally wet land, with impervious subsoil, for the purposes of ordinary agriculture; but some persons have strongly and repeatedly called in question the necessity of draining land when irrigated with sewage ; and the two farms at Tunbridge Wells, to a great extent, and more especially the Reigate farm at Earlswood, have been actually laid out for sewage irrigation on what may be called the 'saturation' principal; so that it appears to the Committee desirable to call attention to the fact, that if drainage is necessary where no water is artificially applied to the soil, it cannot be *less* necessary after an addition to the rainfall of 100 or 200 per cent. But a comparison of the analysis of different samples of effluent waters which have been taken by the Committee from open ditches into which effluent water was overflowing off saturated land, and from subsoil drains into which effluent water was intermittently percolating through several feet of soil, suggests grave doubts whether effluent water ought ever to be permitted to escape before it has percolated through the soil."

There are some other plans that have been suggested for the distribution of sewage,—one is by means of the ordinary agricultural drainage. That is obviously perfectly absurd. It is impossible to imagine that sewage could be purified by turning it into agricultural drains below the roots of the plants. Another plan, very much advocated, is to conduct it in pipes just out of reach of the plow, and then to distribute it over the land and plants by means of hose and jet. The thing that is immensely against this, is the expense from the enormous amount of labor ; also, that if sewage farms are not to be (and they certainly need not be) nuisances, it is not by squirting the sewage about that we shall attain that object. As to the flood plan which is pursued at Milan, I do not think I need say more than two words about it. It is perfectly plain that we don't want to make sewage marshes, at any rate here. There, at Milan, the water meadows cause ordinary marsh fevers. They do not cause any of the diseases that it was expected sewage farms would cause ; they do not favor in any way such diseases as cholera or typhoid fever, which diseases are spread to a very great extent by means of the intestinal evacuations of those suffering from them. I think perhaps as I have gone into the question of public health, I may just say the word or two that I have to say upon that point. I do not think you would find a single case definitely proved against sewage farms — badly conducted as many of them are—where they have been injurious to health, except in this way. If water that contains poison from sewage that has flown over the land and that has not percolated through it, if that water is drunk, it is very likely poisonous. That has been the case once or twice. The farms have not caused by noxious emanations any injury to the health of neighborhoods where they have been placed. I do not think there is the slightest evidence to show that. Then about the water that passes from them. It is said that in certain cases it has poisoned the wells in the neighborhood. There is no evidence of anything of the sort. Where these cases have been inquired into, it has been found that the wells have not been poisoned from the sewage farms, but by foul matters from perfectly different sources. It is true that in one or two cases in which the water which had passed over the sewage-meadows was drunk, a certain number of people got typhoid fever. In sewage-meadows where surface action is relied upon there is considerable danger that the overflow of the channels should be mistaken for fresh water. You understand that when a sewage farm is a large filter, and the effluent water is collected in subsoil drains, this water is perfectly fit to drink. I can tell you of a sewage farm where it is usually drunk by the workmen. There is a well on that particular farm, the water of which is excellent, and there is no reason why it should not be so. Our own drinking water in London, and in most large towns, has got purified from all kinds of impurities by passage through gravel and sand. Dr. Angus Smith tells us that we could not drink rain-water if collected from the clouds anywhere near to large towns ; it would be too foul, and would have to be passed through soil in order to be purified.

Well now, the last point that I have to notice in connexion with the public health is the alleged danger from the spread of entozoic diseases by means of sewage irrigation. Dr. Cobbold, the great authority upon entozoa, thinks that if sewage farms are spread much over the land, we shall have more of entozoic diseases, and that deaths from them will become much more frequent, and he even suggested that an entozon which is very fatal in some parts of Northern Africa (the Bilharzia hæmatobia) might become prevalent in this country. But that entozoon is, in the first place, prevalent in those countries especially during the hot seasons. In the second place, we know next to nothing about the different stages it goes through, during which it no doubt inhabits different animals (snails, &c.); and lastly, Dr. Cobbold himself has shown that the larvæ of this parasite cannot live in impure water. So we may dismiss that at once. Then with regard to ordinary entozoa. In the first place, there is no sort of evidence whatever to show that they have been spread in cattle at farms where irrigation has been going on for 200 years. Professor Christion has distinctly stated that he has never been able to trace entozoic disease to the Craigentinny meadows near Edinburgh, neither is there evidence that this has been done anywhere else. It is very easy to say that the eggs of the entozoa are in the sewage when it is carried on to the land, and that the larvæ will be developed as soon as the plants are eaten by animals. In the first place, you must know that it is necessary that these eggs should be living and fertilized too, and they have the smallest chance of living that anything can possibly have by the time they get with the sewage to the land, for they have a considerable distance to go before they get to the farm; they are tossed about in an alkaline liquid, their natural habitat being acid excretions; they are turned on to the ground and taken down into the soil with the water. However, to prevent any apprehension on this score the simplest thing is to have the grass cut and carried to the stalls, and not to graze animals upon it. Many of the best irrigationists insist upon this.

Some investigations of this matter were made by the British Association Committee.

"An ox which had been fed for the previous 22 months entirely on sewage grown produce" was slaughtered and carcass examined by Dr. Cobbold, Professor Marshall and myself; no trace of any entozoic disease was found in it, although most carefully looked for. Dr. Cobbold suggests several reasons for this result, and one of these is the freedom of the sewage farm from snails and insects, in the bodies of which many of these entozoa go through different stages of their existence; it seems, therefore, that the sewage kills those creatures which are necessary for the existence of these entozoa in their different stages.

Dr. Cobbold also examined under the microscope portions of "flaky vegetable tufts," collected from the sides of the minor sewage carriers, and found that, although they contained animal as well as vegetable life, they contained "no ova of any true entozoon." So that you see that as far as we have got positive evidence it is entirely against the theory that entozoic diseases are spread in cattle, and from them down, by means of sewage irrigation.

Now a few words about the crops. The most suitable crop for sewage is Italian rye grass. This plant will take up a very large amount of sewage. If you read the reports of the sewage of Towns Commissioners you will find the results of experiments upon the amounts of meadow grass grown with different quantities of sewage.

There was an average increase of about 4 tons of grass for each thousand tons of sewage applied per acre: the maximum amount of the latter being about 9,000 tons per acre per annum. The largest amount was about 33 tons of green grass per acre in one year, and 37 tons in another. Some of the land was not supplied with sewage at all ; other parts with 3,000, 6,000, and 9,000 tons per acre per annum. The increase of produce was much greater with the first 3,000 tons of sewage than it was when the amount was increased from 3,000 to 6,000 tons, and more from 3,000 to 6,000 than from 6,000 to 9,000. So that the increased amount of sewage did not produce a proportionately increased amount of produce. The increase of produce per 1,000 tons of sewage was when 3,000 tons were applied about 5 tons of green grass, when 6,000 tons were applied 4 tons 2½ cwt., and when 9,000 tons were applied 3 tons 3¼ cwt. And the results given by Italian rye grass showed about the same increase of produce. It was also found that an earlier cut of green grass could be obtained by means of sewage irrigation.

Experiments were made about the quality as well as about the quantity, and it was found that the grass contained a smaller actual amount of dry solid matters when grown with sewage, but was richer in nitrogen, and was, in fact, more readily assimilable — more milk could be got from it.

The main result of irrigation farms must be the feeding of cattle and the production of milk. The sewage is turned into Italian rye grass, and is returned to the town from which the sewage has come as milk, butter, cheese and beef.

Then, if Italian rye grass can be grown, every grass and almost anything else can as well. You will see that denied even to this day, in spite of the fact that almost everything else has been grown with it. These different plants can be grown upon land which is absolutely and perfectly valueless in an agricultural point of view without sewage, even upon blowing sea sand, and you can see in many parts of England excellent crops now growing by means of the use of this rich manure. Cereals can be grown perfectly well with considerable returns. In 1868 and 1869 (at Lodge Farm, Barking), wheat, winter oats, rye and cabbages were grown. In 1868 wheat was grown on a slope of shingle. It had two dressings of sewage equal to 450 or 500 tons in all. The results were 5 qr. 3 bush., as against 3 qr. 5 bush. without sewage, with 4½ loads of straw, as against 3 to the acre. The winter oats yielded 8 qr. of corn, with three loads of straw to the acre. Among other vegetables must be especially mentioned beet-root. From experiments which have been made there seems very little doubt that beet-root can be grown for the production of sugar in almost any quantity. Professor Voelcker has analyzed some of the beet-roots grown on sewage, and they gave 13.19 per cent. of sugar, while the beet-roots from Holland, Suffolk and Scotland, only gave from nine to ten per cent. of it at the outside.

Well, now a word or two about the times when you don't want sewage on the land. There may be times when you don't want it at all—times when the sewage is too dilute, and the land is very wet, as during heavy floods ; and this is a very strong argument for keeping the drainage water, properly so called, out of the sewage, the utilization of sewage is thereby rendered very much easier. The best dilution for sewage is when it represents 25 to 30 gallons per head of the population. If you keep the drainage water as much as possible separate, you can always turn it into it as a diluent when more water is wanted.

The amount of sewage required per acre varies much with different crops and with different soils, but it is usually considered that the sewage of from 35 to 40 persons is sufficient per acre on the average, although in many instances much more than that is applied.

On every sewage farm there should be a piece of fallow land to be used as a filter, not with the view of any great return, but simply with a view to purifying the sewage whether the crop on that particular land happens to want it or not, when it is not wanted on any other part of the farm.

You see, then, that intermittent downward filtration through soil and irrigation farming, with passage of the liquid through the soil, are the only means at present known for purifying sewage, and these may be well continued, with some deodorizing process, which will prevent the sludge in the tanks from being offensive, except where the tanks are in the open country, when this is hardly necessary ; and you see also that these processes in themselves are in no way injurious to the health of the neighborhood where they are carried on ; one of them, irrigation farming, with the condition mentioned above, also affords the only method known by which the valuable manurial ingredients dissolved in sewage can be utilized—can be turned into wholesome food for man and beast; and it is therefore for you in those parts of the world in which you may be stationed, and where you will have to advise on such matters, to use your influence in obtaining the adoption of the water-carriage system, as before described, in connection with a properly carried out plan of irrigation farming.

The removal of waste matters is the first thing to consider, their utilization the second ; where you have both, there you are best able to compete with disease and death.

*** Any book in this Catalogue sent free by mail on receipt of price.*

VALUABLE

# SCIENTIFIC BOOKS,

PUBLISHED BY

## D. VAN NOSTRAND,

23 MURRAY STREET AND 27 WARREN STREET,

NEW YORK.

---

FRANCIS. Lowell Hydraulic Experiments, being a selection from Experiments on Hydraulic Motors, on the Flow of Water over Weirs, in Open Canals of Uniform Rectangular Section, and through submerged Orifices and diverging Tubes. Made at Lowell, Massachusetts. By James B. Francis, C. E. 2d edition, revised and enlarged, with many new experiments, and illustrated with twenty-three copperplate engravings. 1 vol. 4to, cloth........................$15 00

ROEBLING (J. A.) Long and Short Span Railway Bridges. By John A. Roebling, C. E. Illustrated with large copperplate engravings of plans and views. Imperial folio, cloth.............................. 25 00

CLARKE (T. C.) Description of the Iron Railway Bridge over the Mississippi River, at Quincy, Illinois. Thomas Curtis Clarke, Chief Engineer. Illustrated with 21 lithographed plans. 1 vol. 4to, cloth.............................................. 7 50

TUNNER (P.) A Treatise on Roll-Turning for the Manufacture of Iron. By Peter Tunner. Translated and adapted by John B. Pearse, of the Pennylvania Steel Works, with numerous engravings wood cuts and folio atlas of plates.................. $10 00

ISHERWOOD (B. F.) Engineering Precedents for Steam Machinery. Arranged in the most practical and useful manner for Engineers. By B. F. Isherwood, Civil Engineer, U. S. Navy. With Illustrations. Two volumes in one. 8vo, cloth.......... $2 50

BAUERMAN. Treatise on the Metallurgy of Iron, containing outlines of the History of Iron Manufacture, methods of Assay, and analysis of Iron Ores, processes of manufacture of Iron and Steel, etc , etc. By H. Bauerman. First American edition. Revised and enlarged, with an Appendix on the Martin Process for making Steel, from the report of Abram S. Hewitt. Illustrated with numerous wood engravings. 12mo, cloth.................................... 2 00

CAMPIN on the Construction of Iron Roofs. By Francis Campin. 8vo, with plates, cloth.......... 2 00

COLLINS. The Private Book of Useful Alloys and Memoranda for Goldsmiths, Jewellers, &c. By James E. Collins. 18mo, cloth.................... 75

CIPHER AND SECRET LETTER AND TELEGRAPHIC CODE, with Hogg's Improvements. The most perfect secret code ever invented or discovered. Impossible to read without the key. By C. S. Larrabee. 18mo, cloth. .................... 1 00

COLBURN. The Gas Works of London. By Zerah Colburn, C. E. 1 vol 12mo, boards................ 60

CRAIG (B. F.) Weights and Measures. An account of the Decimal System, with Tables of Conversion for Commercial and Scientific Uses. By B. F. Craig, M.D. 1 vol. square 32mo, limp cloth.... ......... 50

NUGENT. Treatise on Optics; or, Light and Sight, theoretically and practically treated; with the application to Fine Art and Industrial Pursuits. By E. Nugent. With one hundred and three illustrations. 12mo, cloth.................................... 2 00

FREE HAND DRAWING. A Guide to Ornamental Figure and Landscape Drawing. By an Art Student. 18mo, boards,.............................. 50

HOWARD. Earthwork Mensuration on the Basis of the Prismoidal Formulae. Containing simple and labor-saving method of obtaining Prismoidal contents directly from End Areas. Illustrated by Examples, and accompanied by Plain Rules for Practical Uses. By Conway R. Howard, C. E., Richmond, Va. Illustrated, 8vo, cloth............ .. ............ 1 50

GRUNER. The Manufacture of Steel. By M. L. Gruner. Translated from the French, by Lenox Smith, with an appendix on the Bessamer process in the United States, by the translator. Illustrated by Lithographed drawings and wood cuts. 8vo, cloth.. 3 50

AUCHINCLOSS. Link and Valve Motions Simplified. Illustrated with 37 wood-cuts, and 21 lithographic plates, together with a Travel Scale, and numerous useful Tables. By W. S. Auchincloss. 8vo, cloth.. 3 00

VAN BUREN. Investigations of Formulas, for the strength of the Iron parts of Steam Machinery. By J. D. Van Buren, Jr., C. E. Illustrated, 8vo, cloth. 2 00

JOYNSON. Designing and Construction of Machine Gearing. Illustrated, 8vo, cloth.................. 2 00

GILLMORE. Coignet Beton and other Artificial Stone. By Q. A. Gillmore, Major U. S. Corps Engineers. 9 plates, views, &c. 8vo, cloth.................... 2 50

SAELTZER. Treattse on Acoustics in connection with Ventilation. By Alexander Saeltzer, Architect. 12mo, cloth.................................... 2 00

THE EARTH'S CRUST. A handy Outline of Geology. By David Page. Illustrated, 18mo, cloth.... 75

DICTIONARY of Manufactures, Mining, Machinery, and the Industrial Arts. By George Dodd. 12mo, cloth............................................ 2 00

BOW. A Treatise on Bracing, with its application to Bridges and other Structures of Wood or Iron. By Robert Henry Bow, C. E. 156 illustrations, 8vo, cloth.......................................... 1 50

GILLMORE (Gen. Q. A.) Treatise on Limes, Hydraulic Cements, and Mortars. Papers on Practical Engineering, U. S. Engineer Department, No. 9, containing Reports of numerous Experiments conducted in New York City, during the years 1858 to 1861, inclusive. By Q. A. Gillmore, Bvt. Maj -Gen., U. S. A., Major, Corps of Engineers. With numerous illustrations. 1 vol, 8vo, cloth.............. $4 00

HARRISON. The Mechanic's Tool Book, with Practical Rules and Suggestions for Use of Machinists, Iron Workers, and others. By W. B. Harrison, associate editor of the "American Artisan." Illustrated with 44 engravings. 12mo, cloth........... 1 50

HENRICI (Olaus). Skeleton Structures, especially in their application to the Building of Steel and Iron Bridges. By Olaus Henrici. With folding plates and diagrams. 1 vol. 8vo, cloth.................... 1 50

HEWSON (Wm.) Principles and Practice of Embanking Lands from River Floods, as applied to the Levees of the Mississippi. By William Hewson, Civil Engineer. 1 vol. 8vo, cloth........................ 2 00

HOLLEY (A. L.) Railway Practice. American and European Railway Practice, in the economical Generation of Steam, including the Materials and Construction of Coal-burning Boilers, Combustion, the Variable Blast, Vaporization, Circulation, Superheating, Supplying and Heating Feed-water, etc., and the Adaptation of Wood and Coke-burning Engines to Coal-burning; and in Permanent Way, including Road-bed, Sleepers, Rails, Joint-fastenings, Street Railways, etc., etc. By Alexander L. Holley, B. P. With 77 lithographed plates. 1 vol. folio, cloth.... 12 00

KING (W. H.) Lessons and Practical Notes on Steam, the Steam Engine, Propellers, etc., etc., for Young Marine Engineers, Students, and others. By the late W. H. King, U. S. Navy. Revised by Chief Engineer J. W. King, U. S. Navy. Twelfth edition, enlarged. 8vo, cloth. ............... .......... 2 00

MINIFIE (Wm.) Mechanical Drawing. A Text-Book of Geometrical Drawing for the use of Mechanics

www.ingramcontent.com/pod-product-compliance
Lightning Source LLC
LaVergne TN
LVHW021417110826
845150LV00007B/1971

* 9 7 8 1 4 2 5 5 0 9 8 1 1 *